21st Century Junior Library

# DISCOVER THE PLATEOSAURUS

Our Prehistoric World: Dinosaurs

Josh Gregory

Published in the United States of America by:

Cherry Lake Press
2395 South Huron Parkway, Suite 200, Ann Arbor, Michigan 48104
www.cherrylakepress.com

Content Adviser: Gregory M. Erickson, PhD, Dinosaur Paleontologist, Department of Biological Science, Florida State University, Tallahassee, Florida

Reading Adviser: Beth Walker Gambro, MS, Ed., Reading Consultant, Yorkville, IL

Photo and Illustration Credits: © Daniel Eskridge/Shutterstock, cover, title page; © Dotted Yeti/Shutterstock, 5; Eva K./Eva K., CC BY-SA 2.5, via Wikimedia Commons, 6; Daderot, CC0, via Wikimedia Commons, 7; © Dotted Yeti/Shutterstock, 8-9 © rodos studio FERHAT CINAR/Shutterstock, 10; Paul Hermans, CC BY-SA 4.0, via Wikimedia Commons, 13; © Dotted Yeti/Shutterstock, 14; Kritzolina, CC BY-SA 4.0, via Wikimedia Commons, 16-17; © rodos studio FERHAT CINAR/Shutterstock, 18-19; FunkMonk, CC BY-SA 3.0, via Wikimedia Commons, 20



**Cherry Lake Press** is an imprint of Cherry Lake Publishing Group.

Library of Congress Cataloging-in-Publication Data has been filed and is available at catalog.loc.gov.

Cherry Lake Press would like to acknowledge the work of the Partnership for 21st Century Learning, a Network of Battelle for Kids. Please visit Battelle for Kids online for more information.

Printed in the United States of America



Cherry Lake Press

# CONTENTS

# WHAT WAS A PLATEOSAURUS?

Hundreds of millions of years ago, dinosaurs lived on Earth. Among them was the *Plateosaurus*. It was one of the biggest dinosaurs of its time. But it was not a ferocious predator. Instead, it was a peaceful plant eater.

The *Plateosaurus* was larger than many other dinosaurs of its era.

Like all dinosaurs, the *Plateosaurus* is now **extinct**. However, it was once common in what is now Europe. The *Plateosaurus* lived between 229 million and 200 million years ago. We know about it from studying its **fossils**.

This *Plateosaurus* skull can be seen at a museum in Germany.

## Look

You can get a look at *Plateosaurus* fossils for yourself. Many museums have dinosaur fossils. Some even have entire *Plateosaurus* skeletons!

# WHAT DID A PLATEOSAURUS LOOK LIKE?

*The Plateosaurus* had a long neck and tail.

The name *Plateosaurus* means "broad lizard." The *Plateosaurus* was also very long. An adult *Plateosaurus* could grow to be 16 to 33 feet (4.9 to 10 meters) long. However, they were only around 10 feet (3 m) tall.

Scientists believe the *Plateosaurus* walked on two legs.

The *Plateosaurus* had a thick, heavy body. It weighed almost 8,800 pounds (4,000 kilograms). That is heavier than a car! The dinosaur had strong back legs to carry this weight. Its front legs were smaller but still powerful.

The *Plateosaurus* had a long, heavy tail that helped it balance. It also had a tall neck. Its head was very small compared to the rest of its body. It had a long, flat shape. The dinosaur's mouth held its powerful jaw.

## Create!

**Scientists aren't sure what color the *Plateosaurus*'s skin was. Draw your own picture of this dinosaur. Color it in. Be creative with the colors you use!**

The *Plateosaurus*'s long neck was very flexible.

The *Plateosaurus* grazed on branches and raked the ground for food.

# HOW DID A PLATEOSAURUS LIVE?

The *Plateosaurus* was an **herbivore**. This means it only ate plants. The dinosaur's jaw and teeth were perfectly built to eat a lot of leaves. Its long neck allowed it to reach high up into the trees. It could also feed on plants on the ground.

The *Plateosaurus*'s teeth had jagged edges. This helped the dinosaur tear leaves off trees. The *Plateosaurus* also had five fingers on its front feet. This included a thumb with a long claw. This claw helped the dinosaur gather leaves and branches. It might also have helped it dig up tasty roots.

A *Plateosaurus*'s teeth were made for snipping off and shredding plants.

Members of a *Plateosaurus* herd looked out for one another.

The *Plateosaurus* lived in groups called **herds**. A single herd could include hundreds of these dinosaurs. Herd members helped one another watch out for danger. They also likely traveled together in search of plants and water.

## Make a Guess!

**How do scientists know the *Plateosaurus* lived in herds? Remember that we learn about dinosaurs by studying fossils. How could fossils show that a dinosaur lived in groups?**

This *Plateosaurus* was found in Switzerland.

Scientists know a lot about this **prehistoric** animal. There is still a lot to learn, though. They will continue to search for new fossils. They will also study the fossils they already have. Are you interested in learning about dinosaurs? Maybe you will make an important discovery about the *Plateosaurus*!

# GLOSSARY

**extinct** (ek-STINGKT) describing a type of plant or animal that has completely died out

**fossils** (FAH-suhlz) the preserved remains of living things from thousands or millions of years ago

**herbivore** (HUR-buh-vor) an animal that only eats plants

**herds** (HURDZ) large groups of animals that stay together or move together

**prehistoric** (pree-his-TOR-ik) belonging to a time before history was recorded in written form

**skeletons** (SKEL-uh-tuhnz) the frameworks of bones that support and protect the bodies of animals

# FIND OUT MORE

## Books


Dixon, Dougal. *Plateosaurus and Other Desert Dinosaurs.* Minneapolis, MN: Picture Window Books, 2005.

Miller, Henley. *Meet Plateosaurus.* New York, NY: Cavendish Square Publishing, 2015.

Schomp, Virginia. *Plateosaurus and Other Early Long-Necked Plant-Eaters.* New York, NY: Marshall Cavendish Benchmark, 2006.


## Websites

With an adult, learn more online with these suggested searches.

### How Stuff Works—Plateosaurus

Find out more about the *Plateosaurus* and how it lived.

### Walking With Dinosaurs—Plateosaurus

Check out pictures and read more about the *Plateosaurus*.

# INDEX



# ABOUT THE AUTHOR

**Josh Gregory** writes and edits books for kids. He lives in Chicago, Illinois.